AFRICAN CICHLIDS OF LAKE TANGANYIKA

A QUARTERLY

David E. Boruchowitz

yearBOOKS,INC.
Dr. Herbert R. Axelrod,
Founder & Chairman
Neal Pronek
Chief Editor
Dr. Warren E. Burgess
Editor

yearBOOKS are all photo composed, color separated and designed on Scitex equipment in Neptune, N.J. with the following staff:

DIGITAL PRE-PRESS
Michael L. Secord
Supervisor
Robert Onyrscuk
Computer Art
Sherise Buhagiar
Patti Escabi
Cynthia Fleureton
Sandra Taylor Gale
Pat Marotta
Joanne Muzyka

Advertising Sales
George Campbell
Chief
Amy Manning
Director
Jennifer Feidt
Coordinator

©yearBOOKS,Inc.
1 TFH Plaza
Neptune, N.J. 07753
Completely manufactured in
Neptune, N.J.
USA

The Earth split 10,000,000 to 20,000,000 years ago and formed a great rift valley in the middle of Africa. This rift began to fill up with rainwater and other runoffs, and the great African rift lakes were born. The bottoms and sides of these lakes were usually strewn with rocks which, over the years, gradually dissolved making the water in the lakes hard and alkaline.

Biologically, the few fishes and other aquatic animals had little, if any, new genetic material available, so they interbred and each species searched for an ecological niche that enabled it to eat and be protected. Spawning techniques also changed with the evolution of the species of fishes. Most of the fishes were cichlids and their interesting colors and breeding habits made them a favorite with hobbyists. This yearBOOK will introduce you to the lake and its cichlid inhabitants. It will also help you in preparing their aquarium.

What are yearBOOKS?

Even though it took nature up to 20,000,000 years to develop the present living fauna in Lake Tanganyika, the exploration of the lake is essentially taking place NOW! Every year new species and new color morphs are being discovered. Expert hobbyists are learning how to breed the cichlids from Lake Tanganyika and to raise their fry. Unfortunately, by the time this information finds its way into a book, the information could be a few years old. In order to get up-to-date information out to the general public, yearBOOKS,Inc. is publishing books in magazine format. This means lower prices because of the support of generous advertisers and the less expensive way magazines are manufactured.

Because this information is so valuable, a hardcover edition, sewn for longevity, has also been made available.

The Foto-Glaze™ used to enhance the beauty of the color photographs is protected by U.S. Patent 5,249,878.

Contents

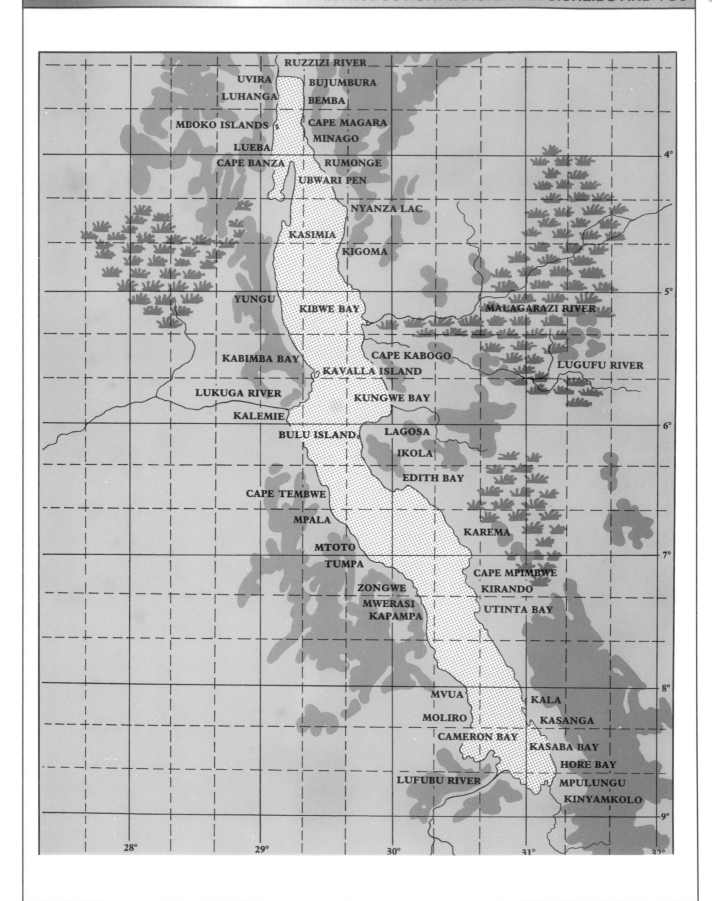

RUZZIZI RIVER
UVIRA
LUHANGA
BUJUMBURA
BEMBA
MBOKO ISLANDS
CAPE MAGARA
MINAGO
LUEBA
CAPE BANZA
RUMONGE
UBWARI PEN
NYANZA LAC
KASIMIA
KIGOMA
YUNGU
KIBWE BAY
MALAGARAZI RIVER
KABIMBA BAY
CAPE KABOGO
LUGUFU RIVER
KAVALLA ISLAND
LUKUGA RIVER
KUNGWE BAY
KALEMIE
BULU ISLAND
LAGOSA
IKOLA
EDITH BAY
CAPE TEMBWE
MPALA
KAREMA
MTOTO
TUMPA
CAPE MPIMBWE
ZONGWE
KIRANDO
MWERASI
KAPAMPA
UTINTA BAY
MVUA
KALA
MOLIRO
KASANGA
CAMERON BAY
KASABA BAY
HORE BAY
LUFUBU RIVER
MPULUNGU
KINYAMKOLO

4°
5°
6°
7°
8°
9°

28°   29°   30°   31°   32°

Introduction:
Tanganyika Cichlids and You

So, you want to keep Tanganyika cichlids? Great! You're starting the right way, by reading about these fishes and how to care for them. Half of the known cichlids in the world live in Africa, and a great number of them live in the rift lakes of Malawi and Tanganyika. This book provides an overview of the fishes of Lake Tanganyika and an introduction to their husbandry.

Since these cichlids are not really beginners' fishes, we assume that you've at least maintained a tank or two for a while and probably bred some livebearers and maybe some egglayers. Or you've kept fishes for a longer period of time but have not yet tried the Africans. Maybe you've already been bitten by the cichlid bug and have kept and bred some Neotropical species and now want to learn about their Old World cousins. Perhaps you've been charmed by some you've seen in your dealer's tanks, or in a friend's, or maybe you just want to find out what all the furor is about. In any case, be warned: "African Rift Fever" is contagious! Once you start with these fascinating little beauties, you'll probably be hooked and start eyeing odd corners of

Above: One of the newly discovered *Lamprologus* is this *L. gracilis*.
Below: Another discovery from Lake Tanganyika is the cichlid *Tanganicodus irsacae. Both photos by MP & C Piednoir, Aqua Press.*

LAKE TANGANYIKA

Above: The typical habitat of *Tropheus duboisi*. The *duboisi* shown here are young. Photo by Ad Konings. *Below:* The entire shoreline of Lake Tanganyika is dotted with habitats suitable for rock-dwelling cichlids. Since the level rises and falls, what we see here is absolutely typical of an underwater scene. Photo by Büscher.

Understanding and caring for Tanganyika's cichlids requires first of all a knowledge of their natural habitat. The large African Rift Valley was formed where two plates of the earth's crust pulled away from each other, creating a huge trench. The rift lakes are in effect large drainage pits. For millennia they have filled with water, giving it up through evaporation. Like infant oceans, they have in this way developed a high concentration of dissolved minerals, and while not salty in the sense of containing an excess of dissolved sodium chloride, their water, though still "fresh," contains an abundance of dissolved minerals. In Lake Tanganyika, the

concentration is so great that at a depth of more than fifteen feet carbonate salts of sodium, calcium, and magnesium precipitate out into a crust of sharp crystals on rock surfaces.

Also like seas, these lakes are immense temperature buffers, with the water temperature highly unresponsive to temporary weather conditions. Coupled with the equatorial climate, this produces almost no variation in temperature by season, by location, or by time of day. There are even substantial waves on these lakes, which, coupled with alternating sandy and rocky

Above: A school of *Lamprologus brichardi* taken at a depth of about 30 feet. This photo was taken by Pierre Brichard. *Below:* Life along Lake Tanganyika is primitive compared to life in other parts of the world. Yet the clear air, perfect temperature of about 75° F. all year round and crystal clear water and magnificent fishes, make up for the lack of running water and electricity. Photo by Büscher.

beaches, provide a very marine-like panorama. There aren't any sharks, but scuba divers, who are used to those cartilaginous coral reef menaces, should find enough excitement in the twenty-foot long Nile crocodiles, irascible hippos, and deadly water cobras.

These divers might do a double take on the first dive, for in these lakes rocky "reefs" are characteristic, and the cichlids resemble marine reef fishes in their variety of bright colors and patterns. Similar evolutionary forces have operated on these fish populations, since the reefs and their rubble zones provide habitable "islands" in a biological desert of deep water which lacks the diversity of food choices found along the coastlines. While some species do make a home in the deeper water, living off the abundant plankton blooms, they are always close to the bustling communities of the rocky shallows, where the crystal clear water and intense tropical sun provide a luxurious growth of algae, along with a profusion of invertebrates that graze on the vegetation-covered rocks. Along with the plankton, this biocover is the foundation of the Tanganyika food chain.

Almost all the fish species are associated in some way with the copious boulders, stones, and rubble of the lakes. They provide refuge, breeding

When Dr. Herbert R. Axelrod and Pierre Brichard discovered the aquarium potential of the fishes in Lake Tanganyika, this was the density of the fishes on the rocks outside Bujumbura, Burundi. Photo by Dr. Herbert R. Axelrod.

sites, grazing or foraging zones, and areas for establishing territories. Even the deeper water species live in association with rocks on the bottom, and many come into the rocky shallows to breed. Most cichlids live very close to the coastline, with different species' ranges extending from the wave-tossed shores out along the underwater jumble of rocks in a sometimes precipitous slope. Most are found in the rocks, while some inhabit the sandier zones, but none live in the wide

Photographed in 1972, the rocks in certain areas of Lake Tanganyika were covered with soft sponges and algae upon which the fishes fed. Photo by Dr. Herbert R. Axelrod.

open waters of the middle of the lake. In fact, the deep water areas far from shore represent the only ecological niche cichlids have not really filled.

The history of these lakes is a natural biological experiment. When the lakes were first formed, cichlids were probably no more numerous in their waters than they now are elsewhere in the lakes and rivers of Africa. Today, however, they are the predominant higher life form of the rift lakes, with hundreds of descendant

A typical reef site in Lake Tanganyika. You should set up your aquarium like this, with lots of hiding places for the fishes. Photo by Dr. Herbert R. Axelrod.

species. In fact, African rift lake cichlids account for a substantial percentage of all cichlids on earth. Had Charles Darwin known about the fauna of the African rift lakes, he likely never would have bothered with his finches, for the variety and specializations of cichlid species in these lakes is an evolutionary case study *par excellence*.

The main factors in speciation are time, isolation or separation, and a variety of potential habitats, and the evolution is driven by competition; the more intense the competition, the greater the speciation. Lake Tanganyika has provided its fishes with all of these.

It is the oldest of the rift lakes, with estimates for its age between 10 and 20 million years. During all this time the water level has varied considerably from that of the present, such that what are now underwater mountain ranges have, in the past, divided the lake into separate bodies of water, when the level was much lower. Most of this vast volume of water is devoid of fish life, however, and the cichlids of the lake are concentrated in the relatively shallow shorelines, which alternate from rocky to sandy and back again.

Geographical isolation has been effected in two ways: actual separation of multiple bodies of water where there is now one, and the separation of populations at various islands and reefs, which, while uninterrupted water stands between them, are effectively cut off from each other, since the rock-hugging fishes will not normally venture across deep, open water. In addition, the oxygen-bearing portion of the water is only the top 300 to 600 feet; the deeper water, which almost never mixes with the surface layers, is anoxic, forming an absolute

A typical fishing village along Lake Tanganyika. The local people are taught how to collect the small cichlids and to keep them alive and healthy. This is their only possibility to earn hard currency and is, therefore, encouraged by their government. However, the local civil wars have disrupted this commerce. Bujumbura, Burundi, the primary site for the export of fishes because it has direct flights to Brussels, has seen many thousands of people killed. When Dr. Axelrod took his first dive into Lake Tanganyika outside of Bujumbura, he saw thousands of human skulls lying on the bottom of the lake.

deterrent to the mixing of populations by migration along the bottom, even if the fishes could stand the water pressures at those depths.

The variety of habitats necessary for diversification of species is provided by the multiple ecological niches of the shorelines. There is not much higher plant life, but the algae-

opportunistic feeders with a smorgasbord of fish, egg, and fry choices. Tanganyikan cichlids have diversified and filled all these niches; there are even species that specialize in stealing and eating scales off neighboring species of fish, and some that have developed large, protrusive mouths, which they use to suck smaller fishes out of rocky crevices.

The various types of cichlids all share the waters of Lake Tanganyika, however, and since water is the largest component of any aquarium, it is important to understand the nature and chemistry of this lake. As the oldest rift lake it is

decide if keeping Tanganyika cichlids is still for you.

A local Belgian resident of Burundi has a pygmy hedgehog as a pet. These small animals are the latest rage in the U.S.A. now. Photo by Dr. Herbert R. Axelrod.

The late Pierre Brichard, who passed away on March 14, 1990 in Burundi. Brichard collected and shipped fishes from Lake Tanganyika in 1972 when he explored the lake with Axelrod. Photo by Dr. Herbert R. Axelrod in Kinshasa, Zaire, June, 1969.

laden rocks provide rich grazing for herbivores, with swarming herds of insects, crustaceans, and other invertebrates for micropredators. The rubble piles provide a multitude of cracks, caves, and crevices for shelter and breeding. The large fish population provides predators and

also the one most full of dissolved salts, with a pH of 8.8 to 9.3, a hardness of over 10 degrees, and a carbonate hardness as high as 18. Providing this type of water is the foremost requirement for keeping these fishes. So, test your tap water, read the following chapter, and

WATER, WATER, WATER

To paraphrase advice from the real estate profession, the three keys to the successful keeping of Tanganyikan cichlids are: water, water, and water! By and large they are not difficult fishes to keep, but they must have water with specific properties, as do, say, discus from South America. What makes them unusual is the type of water they require.

There are two things that make providing proper water for rift lake cichlids generally easier than for discus. (1) Many more people have suitable tap water for the former than for the latter, and (2) since the rift lake cichlids need hard, alkaline water (water with a large quantity of minerals dissolved in it), and the Amazonian cichlids need soft, acid water (water with almost nothing other than gases and some organic compounds dissolved in it), it is much easier to prepare appropriate water for the African cichlids by adding required chemicals than it is to remove unwanted substances for the Neotropical cichlids with reverse osmosis or resin filtration units.

Replicating Tanganyika's Water

Most pH and hardness test kits have good explanations and instructions, and your dealer can help you test your water. Two types of hardness are measured, usually in degrees from 0 up: general hardness, which refers to the total amount of dissolved minerals, and carbonate hardness, which indicates the buffering capacity of the water, that is, its ability to remain at a stable pH.

Tests kits enable the hobbyist to monitor pH, hardness, ammonia levels, etc. These kits are available from your local pet shop. Photo courtesy of Aquarium Pharmaceuticals.

Soft water has few dissolved minerals and little or no buffering capacity, while very hard water has high concentrations of dissolved minerals. The pH (proportion of hydrogen) measurement, which is measured on a scale of 0 to 14, gives the concentration of H^+ ions (acidity) in the water. Neutral water, such as distilled water, has a pH of 7.0, with the strongest acid having a pH of 0.0, and the strongest base (most alkaline) having a pH of 14.0. Most freshwater fishes prefer a pH somewhere between 6.5 and 7.5—but not Tanganyika cichlids.

You can keep Tanganyikan cichlids even if you only have soft acid water at your disposal, though American cichlids of the genus *Symphysodon* or *Apistogramma* would be a lot easier to keep. For the Africans you would have to treat every single drop of water you place into your tanks to raise both its pH and its hardness to acceptable levels. Given the large volume of water needed to set up and maintain most Tanganyikan tanks, this can be quite costly and time-consuming.

If you are fortunate enough to have tap water that tests out with a pH of at least 7.8 or, preferably, higher, and with a total hardness of near 10 degrees and a high carbonate hardness as well, you're all set to get out the hoses and limit conditioning of the water from your tap for removal of chlorine, chloramine, or heavy metals, if necessary. Harder or more alkaline water is not a problem for Tanganyikans, and it is in fact preferable, so if you have a well sunk in limestone, your fishes will feel right at home in your water.

Several commercially available "rift lake salt" mixtures offer an easy way to provide needed minerals to deficient water; make sure that you condition any replacement water in a bucket, mixing the additives thoroughly, then test it to make sure it matches the water already in your tank. It is important to avoid shocking the fishes by adding water of different composition, even if the new water would in itself be preferable. Make any changes gradually.

I go so far as to suggest that if you have minimally acceptable water, such as pH 7.7, you might opt against adding rift lake salts to make the water more ideal for your cichlids. The reason for this is that the most important consideration for the water in your tanks is to keep it there for as short a time as possible! Frequent, constant, religiously observed water changes are essential to the good health and breeding success of these fishes (as they are for almost any fishes), so any obstacle to changing the water can be considered a drawback. If you don't have to test, measure, mix, and retest, you're much more likely to keep up with the changes. Of course, you

still want occasionally to test the tank water and the new water, because any supply can vary, and you want to keep tabs on the chemistry in your aquaria.

Additive salt mixes, the use of soluble rocks, and chemical manipulation of your water are all viable options if you wish to keep Tanganyikans and have only an unsuitable water source, but you must realize from the start that

carefully.

Maintaining Water Chemistry

Water in an aquarium does not necessarily stay at the same pH and hardness it had when it was originally placed in the tank. As water evaporates, only the water molecules escape, leaving an increasing concentration of dissolved minerals behind. You may not like the scaly film on your glassware, or

Water conditioners are crucial when maintaining a natural environment for your cichlid. Photo courtesy of Mardel Laboratories.

keeping up with the maintenance of your tanks is going to be a much more involved chore. Raising cichlids from Lake Tanganyika is easiest if your water is already close to what they require, but it is possible to get your heart's desire if these fishes really appeal to you but you are cursed with water a discus breeder would call a blessing. You will simply have to condition the water

the crusty ring around the water line in your aquarium, but your cichlids benefit from the increased hardness. If you top off the tank between water changes, this also results in a dissolved concentration higher than the original one (though not as high as before you added water), since you replaced a volume of pure (evaporated) water with an equal volume of mineral-laden water.

Views of the shallows (20 feet or less) of Lake Tanganyika showing the rocky areas as well as some of the sandy areas. Photos by Pierre Brichard.

While the normal change in hardness in an aquarium due to evaporation is not great, pH fluctuation can be much more drastic. In an established tank, decomposition is always occurring. Fish wastes, dead plant materials, even an unnoticed dead fish wedged into a cave all add to the bio-load of the tank and drive down the pH. The decomposition depletes oxygen and increases carbon dioxide, which makes the water more acidic. In soft, unbuffered water the drop can be rapid and dangerous.

Water suitable for Tanganyikan cichlids is less susceptible to this type of pH drop due to the buffering capacity of the water, and a high calcium hardness level is the best insurance against pH drop. Also, if you use calcareous rocks or substrates in the tank, they will slowly leach minerals into the water, raising both pH and hardness levels. This happens faster at lower pH's (more acidic), so functional decorations of this type can be important buffers. You've probably heard warnings about using certain types of rocks

as decorations in a freshwater tank—limestone, coral rock, and sea shells are supposed to be reserved for marine aquaria. But remember that Lake Tanganyika is similar to an ocean in its high concentration of dissolved minerals, exactly those minerals that leach

POUCH TREATS UP TO 55 GALLONS

Nitra-Zorb

Removes Ammonia, Nitrite and Nitrate

RECHARGEABLE

For Freshwater Aquarium Use Only

AQUARIUM PHARMACEUTICALS

A number of products have been developed to aid in the removal of dangerous nitrogenous wastes from the aquarium. Photo courtesy of Aquarium Pharmaceuticals.

out of these decorations and make them unsuitable for the average freshwater aquarium. So, especially if you're trying to raise the pH and hardness of your water, these make good, functional decorations for the Tanganyika tank.

In addition, your pet supplier should have a variety of substrates normally sold for salt water

tanks that are appropriate for the Tanganyika aquarium. Dolomite pieces, crushed coral, or ground shells can be found in an assortment of colors and particle dimensions. Sand or small gravel sizes are best, while large chunks should be avoided for two reasons: (1) your cichlids will not be able to practice their natural digging and excavation if the pieces are too big, and (2) the large interstices provide a disaster-waiting-to-happen when uneaten food is trapped in them and begins to decompose.

It would take an impossibly long time for a chunk of limestone dropped into a tank of distilled water to create Tanganyika-like water chemistry, and even the maximized surface area of a bed of coral gravel can only do so much to raise the pH and hardness. These rocks can, however, add appreciably to the buffering capacity of the water, since acids act directly on them, leaching out the beneficial minerals, which at the same time neutralize the acids. They can help marginal water become more ideal and offer important buffering for any Tanganyika tank.

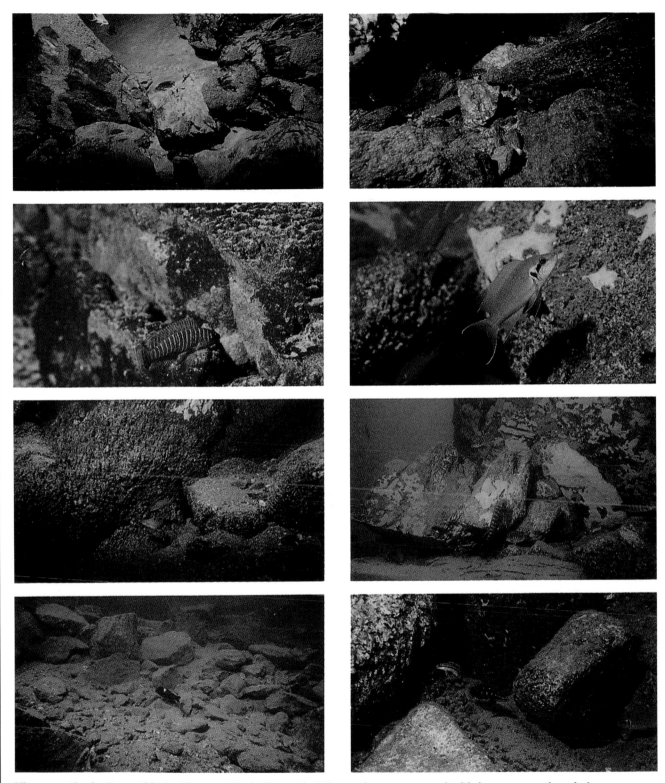

Views on the bottom of Lake Tanganyika indicate that the rocks are covered with heavy growths of algae, sponges and plants. The fishes feed on these growths and their teeth are specially adapted for removing these growths from the rocks. Photos by Pierre Brichard.

CHANGING WATER

What? An entire chapter on water changes? Yes! Frequent partial water changes cannot be stressed too much in fish keeping, and in the Tanganyika tank they are more vital than for many other species, since rift lake cichlids evolved under conditions of pristine water and are not biologically equipped to tolerate very much at all in the way of organic impurities. They thrive in mineral concentrations that would gag most other fishes, but they cannot withstand any buildup in waste products.

To enjoy your Tanganyika "mini-reef" you do not have to concern yourself with the details of water chemistry that give marine reef aquarists nightmares, but you do have to strive for the best possible water conditions, and while all aspects of tank maintenance are important, water changes is the one aspect you must never skimp on. No amount of filtration, for example, can make up for insufficient water changes, but a slightly inadequate filtration system can be made up for by increased water changes. There is nothing mystical about all this; water changes partially make up for the fact that an aquarium is a horribly tiny, closed system, in which fishes are locked up in a small volume of water with no place for toxic and noxious substances to go except swirling around and around with them. Imagine yourself locked in a room with no outlet for any of your waste products! How long would you like to stay in there?

In Lake Tanganyika, the fishes are exposed to their own wastes only for a brief moment after they excrete them, just until the currents carry them off, diluting them in the millions of gallons of lake water. Even if there was no biological breakdown of the wastes going on, which there is, it would take centuries to realize the concentration of wastes that an aquarium can reach in a matter of days or weeks.

The best simulation of the natural environment is a drip/overflow system, where a constant stream of water of the proper temperature and chemistry spills into the tank, and the tank is equipped with an overflow to carry off and discard water at the same rate. This requires some expense for installation, and a good, cheap supply of water of the proper type. If

One side of the Lake is Zaire, the old Belgian Congo. The rebels occupy this area and shoot at anyone coming within rifle range. There are many nice aquarium fishes on the Congolese side of Lake Tanganyika but they will have to await some sort of peace treaty. A 1973 photo by Dr. Herbert R. Axelrod.

Pebble shore with reed growth on Lake Tanganyika.   Rocky shore with boulder zone on Lake Tanganyika.

The rock zone with underwater boulders. Photos by Dr. Herbert R. Axelrod.

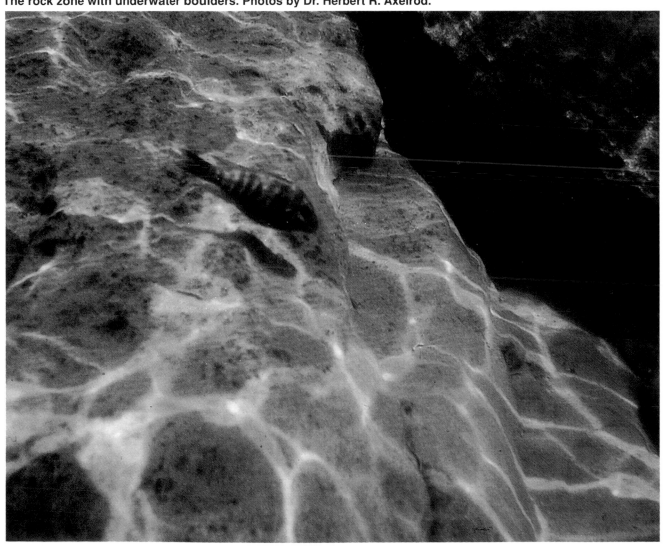

your water supply must be modified for your cichlids, it is possible to design a gravity-fed system so that you can create a reservoir of properly conditioned water which feeds the drip supply lines. If the flow is slow enough the temperature need not be a concern since the aquarium heater will keep up with it, and even modest flow rates add up to a lot of water being changed per day.

This fact lies behind devices that imitate this more sophisticated design and are meant to be used occasionally for a short period of time, then moved from tank to tank. These gadgets have an input hose and a discharge hose. The head houses a spray nozzle, which adds water to the tank, and a water-driven pump, which removes water faster than it enters the tank, so that it can be used to lower the water level. The drain tube regulates the depth of the water by the simple fact that when the water is below the pipe it cannot be sucked out of the tank. This is the theory, and in practice they work quite well, but they have to be watched carefully, something the instructions honestly state. Of course, these devices are not as useful to you if your water supply has to be greatly modified for your tank.

Another water changing option is the single-hose product with a switchable water-driven pump that connects to a faucet. With the pump switched open the water rushing through it creates suction drawing water out of the tank, and it flows down the drain along with the water that

There are products available at your pet shop that will make water changes easier and less messy. Photo courtesy of Aquarium Products.

powers the pump. With sufficient water pressure this device can actually pump water up a flight of stairs, though not very quickly. When you close the pump it ceases to function, and the water from the faucet flows instead through the hose and into the tank.

The time honored, water conserving, but slightly messier alternative still exists of using a siphon hose and buckets to empty and refill the aquarium. It is the only practical method if you have to make serious modifications to your water before putting it into the tank, or if you have no choice but to locate your aquarium too far from the nearest sink.

When you siphon the water out of the tank, use a hose with a gravel tube attached. This rigid tube, of larger diameter than the hose, reduces the suction so that the dirt is sucked out of the gravel without the gravel being drawn through the hose. In addition, the gravel is agitated, which releases trapped waste particles. It is not difficult to time a thorough cleaning of the gravel bottom of a tank to remove the amount of water you wish to change. If you think of gravel cleaning as simply the means by which you remove water for changes, you will be doing your fishes a double favor, as well as saving time.

You can choose a siphon and bucket, water-driven pump, water changer, overflow system, or any other device or procedure to change the water in your Tanganyika tank, but however you do it, do it!

Keep the water changes small, say 10% to 15% at a time, but keep them coming! No particular

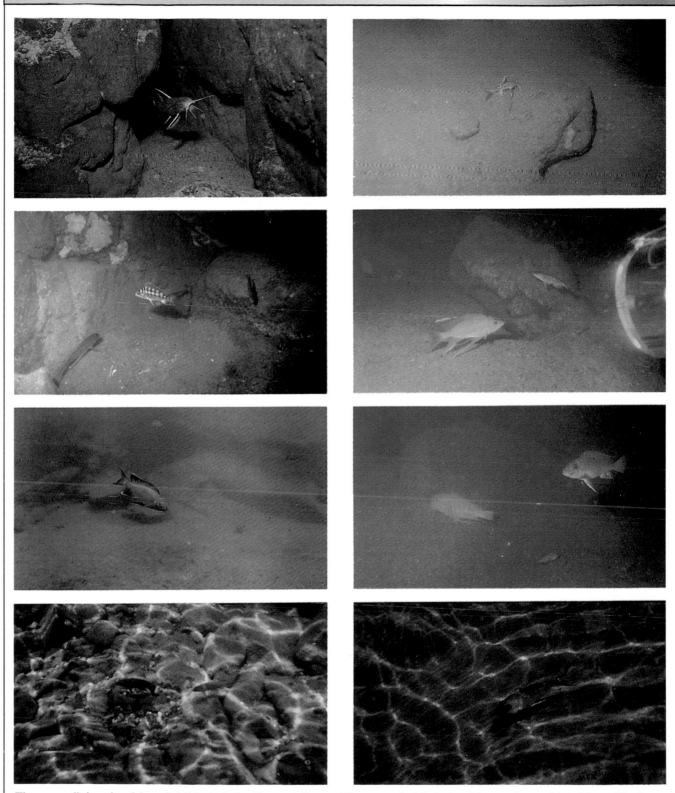

There are fishes besides cichlids in Lake Tanganyika and there are habitats besides rocks and boulders. Photos by Pierre Brichard.

schedule is correct, and the only rule of thumb is: the more the merrier. Daily water changes are fine, though not necessary, but you certainly do not want to draw them out further than once a week at most. Your goal is to keep water chemistry constant and dissolved wastes to an absolute minimum.

There is no other single procedure you can perform that will mean more for the health, longevity, and breeding success of your fishes than regular, frequent, partial water changes.

A magnificent pair of *Altolamprologus compressiceps* photographed in the aquarium by MP & C Piednoir, Aqua Press. In most books this fish is called simply *Lamprologus compressiceps*, but as the *splitters* do their wretched work, they make the groups of related fishes smaller and smaller. A *splitter* looks for differences between fishes in order to split them into new species and genera. A *lumper*, on the other hand, looks for similarities by which he can prove relationships and clump small groups with minor differences.

don't have to don SCUBA gear and join them to keep your wild-caught fish from getting homesick.

Secondly, in order to enjoy their aquaria, all hobbyists compromise on authenticity. Using filtration is not natural, but it not only makes up for other factors that are part of the natural environment, it increases the carrying capacity of your tank. To provide a "natural" fish-per-gallon stocking density, you would have to divide the number of fishes in Lake Tanganyika by the volume of water in the lake. Even if you used just the volume of the upper oxygenated layer, you would find that you could keep very few fishes, if any, even in the largest aquarium. If, on the other hand, you were to measure the number of fishes per cubic foot of water in one of the rocky zones near shore, you might come up with a stocking density too great for any tank. So, "authentic" means different things, depending on the parameters under consideration.

Some hobbyists might refrain from mixing fishes from different microhabitats, such as surf-dwelling species like *Eretmodus* with sandy-bottom inhabitants like *Xenotilapia*, let alone mixing Tanganyika cichlids with Malawi cichlids. But microhabitat is only one parameter of authenticity,

and you should create a Tanganyika microcosm for your own pleasure. Since you are not going to have an absolutely authentic piece of Tanganyika in your home, you can make choices according to your preferences without worrying about breaking some "Rule of Authentic Biotype Reconstruction."

Create the kind of Tanganyikan microcosm that appeals to you. If stocking a biotype resembling a certain microhabitat in the lake with only those species found in that habitat appeals to you, fine. If you prefer a tank of just mouthbrooders, or just cave spawners, also fine. If you stock your tank by choosing *Lamprologus sexfasciatus* because they're striped, and *L. leleupi* because they're yellow, there still is no problem, as long as your choices are informed, and you provide the fishes with appropriate water, hiding places, food, and compatible tank mates.

Once you've decided that Tanganyikans are for you, and you're prepared to provide appropriate conditions, you must decide what species you want and what kind of Tanganyikan tank you are going to have if biotopic issues are important to you. The two major elements to consider in deciding how authentic you wish your "window into

Tanganyika" to be are habitat and stocking density. We'll consider them separately.

Tanganyikan Habitats and Their Inhabitants

There are several distinct habitats in the lake. Beginning at the shoreline, there is the surf zone, where waves break against rock or rock-and-sand shores. The water here is in constant motion, highly oxygenated, and slightly higher in pH. The currents are strong, and the fishes that live here have adaptations enabling them to stay put, most notably a vestigial swim bladder, which makes them less buoyant. Like many dwellers of swift streams, they hop rather than swim, and when not in motion they sink to the bottom. The goby cichlids are natives of this environment, and aquaria for them must take into account their great need for oxygen and their habit of grazing algae from rocks.

There are various rocky and sandy zones in deeper water that are home to grazers like *Tropheus* and *Petrochromis*, as well as the lamprologine cichlids, which are micropredators or predators. While the algal grazers tend to occur in schools over the rocks, the *Lamprologus* and their allies live in and among the rocks, often as pairs, feeding on invertebrates and fish fry and eggs. The

Underwater views of Lake Tanganyika as photographed by Pierre Brichard in 1975.

Stocking Densities

Aside from physical limitations as to how many fishes a given aquarium can safely house, the stocking rate will be in part determined by your choice of species. For many species, aquarists can use high population densities to minimize aggression. This perhaps counter-intuitive practice works because the natural aggressiveness of the fishes causes injury and death only as an artifact of aquarium life.

In the wild, most of these fishes are urbanites. In contrast to, say, their country cousins the Amazonian cichlids, Tanganyikan cichlids must eke out a living with hundreds if not thousands of competitors in densely packed populations associated with the rocky zones. To take the analogy one step further, an urban cabby would be judged rude or even belligerent on the streets of a rural village, while his behavior would be normal self-preservation, and quite unremarkable, in the city. Totally uncalled for on the village main street, the aggressive driving habits of the cab driver enable him to negotiate the bustle of city streets without losing his share of opportunities to make his way through the traffic. In the same way, the highly aggressive habits of many Tanganyikan cichlids are part of a repertoire of behavior that is successful and safe in the natural

A pathetic sight is a group of rock-dwelling Tanganyikan cichlids in a tank without rocks and hiding places. The poor fishes lay on the bottom in a dazed and disoriented condition. Photo by Edward Taylor. The fishes are *Lamprologus kendalli* and *Eretmodus cyanostictus.*

environment. Like a timid driver stuck at the curb against a constant flow of traffic, a cichlid would never eat or reproduce if it didn't fight for its share. If he were too slow to react, several interlopers would be in his territory before he even made any kind of response. Even with this excess of aggression, injury is uncommon because a belligerent fish doesn't take all of its fury out on any one fish, since there is always someone else encroaching on its territory to redirect its attack away from the original victim.

The problem comes in an aquarium. Throw just two rivals into the glass confines of a tank and all the fury of the dominant fish's attack has only one target. In addition, the weaker fish has no place to run. With no other targets to bleed off some of the attack, and with no large expanse of rocky bottom into which to escape, such a battle often ends in tragedy. In the unnatural environment of an aquarium, the instinctive behaviors that ensure the safety of the fish and that actually serve to keep a lively, dynamic peace, instead spell stress, injury, and death.

To prevent dangerous aggression we must remove all chances of attack through extremely minimal stocking rates, or we have to use such high population densities that the risks of aggression are minimized. This alternative mimics the natural situation, and even though the fishes are not able to set up tightly-packed but separate territories, the dilution of aggression among so many individuals keeps a tense peace.

Your choice will depend

partly on the species you are dealing with, and whether you want your fish to breed. Monogamous non-mouthbrooding species need a relatively large, permanent territory to breed, and they are usually ferocious in protecting it against both conspecifics and fishes of other species. If you stock heavily, as soon as the fish pair off and try to establish territories, you are likely to have a disaster. This does not mean that you are limited to keeping one pair of fish per tank, however.

First of all, most species of *Lamprologus*, for example, protect territories of about 24 inches square, so a 48-inch-long tank could house two pairs. Also, you can keep a pair in a tank with a denser population of a polygamous mouthbrooder, which need only temporary spawning territories and will spend most of the rest of the time chasing each other around.

Another way of increasing the number of fishes you can keep

together is to combine pairs of monogamous fish of different species. You have to be careful, since these pairs defend against all other fishes, but they are much more aggressive in driving off members of their

A *Lamprologus elongatus* protecting his fry which are hiding in the hole underneath him. Photo by Glen Scott Axelrod.

own or similar species. I have, for example, maintained a 48-inch-long rift lake community that contained several polygamous species, plus two pairs of *L. brichardi* and two pairs of *Julidochromis dickfeldi*. The *L. brichardi*, which were raised together, set up adjacent territories under two different rock piles. The *J. dickfeldi* took over opposite ends of the tank, in one case establishing a "second story" territory higher up in the rock pile that one of the

L. brichardi pairs was using. There was a lot of activity, but things remained peaceful enough, and some fry of each survived. One drawback of this type of setup is that you can wind up with hybrid babies. Most rift lake hobbyists frown on allowing species to hybridize, and many extend this restriction to different subspecies of the same species as well. The motivation here is to maintain the original, natural types. As a general rule the beginner should try to avoid hybridization. Most hybrids lack the beauty of the parents anyway.

With the mouthbrooders you can use the alternative approach. While a single pair of cave spawners will often be very successful alone in a smaller aquarium, a single pair of, say, *Tropheus*, would be very difficult to maintain safely. The best way to house such species is to keep a group of 10 to 20 in a large aquarium in order to diffuse the aggression among many individuals.

BESIDES WATER, WHAT? TANGANYIKAN HARDWARE

A male *Tropheus moorii* guarding the entrance to his home in an inverted flowerpot. The spots in the male's anal fin are incorrectly referred to as *egg dummy* spots. Photo by B. Allen.

While water is extremely important, it is obviously not sufficient for a healthy tank. You must also choose the aquarium itself and the proper equipment to go with it. From your fishes' point of view, three factors must be considered: temperature, filtration, and furnishings. The last of these, reconsidered as decoration, may also be a factor for you, and lighting can be significant for both you and your cichlids. In this chapter I will discuss choosing, purchasing, and setting up the hardware for your tank.

The Aquarium

Obviously, the most important piece of hardware is the aquarium itself. You must consider both size and shape. Since most cichlids stay fairly near the bottom, and since many of them are territorial, some fiercely so, the area of the bottom of the tank is the single most important size factor in a Tanganyika tank.

A long, wide tank is much preferable to a high, narrow one of the same capacity. A "15 high" has a bottom measuring 20 in. by 10 in., with an area of less than 1.4 square feet (200 in² = 1.38 ft²). The floor of a standard 15-gallon aquarium measures 24 in. by 12 in., for an area of a full two square feet! (288 in² = 2.0 ft²). Similarly, a 100-gallon tank (72 by 18, 18 inches high) and a 135-gallon tank (72 by 18, 24 inches high) have the same bottom area of nine square feet.

The same argument applies to surface area, which is also a population-limiting factor, since it is only at the surface that gas exchange—oxygen in, carbon dioxide out—can occur. Thus, a tank of greater surface and bottom area will provide both more aeration and more territory space, increasing the number of fishes that can be kept. Likewise, two tanks of different size but the same surface area, such as the 100-gallon and the 135-gallon tanks, have virtually the same carrying capacity.

Neither a glass nor an acrylic tank is preferable. Glass is cheaper, heavier, hard to scratch, less insulating, and slightly less transparent. That means, of course, that acrylic is more expensive, lighter, easier to scratch, better insulating, and extra clear. Make your choice after examining the multitude of options available at the larger aquarium retailers. You can even order a custom aquarium, so if money is no object you can turn your living room into a floor-to-ceiling, wall-to-wall slice of Tanganyika.

While that is unlikely, you may wind up choosing a fairly large tank, 70 or more gallons in size. Remember that even though such a large aquarium has substantial weight, the water itself weighs much more. A *rough* rule of thumb for tank, water, and gravel weight in normally-equipped tanks is a total of ten pounds or more per gallon. Notice that that beautiful 135-gallon tank is

going to strain your floor to the tune of close to three-quarters of a ton! Unless you are placing your tanks on the concrete floor of a basement, always check into the load-bearing capacity of the spot you have picked, and take appropriate corrective measures, such as installing jacks, telescoping support posts, or extra joists, to shore up the floor. If you do not feel competent to judge and correct the strength of a floor, hire a professional to do the job.

a competent architectural engineer, be extremely cautious in using anything other than a commercially-made stand designed specifically for the size and shape aquarium you are setting up. Few other common household objects even approach the density of filled aquaria, and common household structures and pieces of furniture will not usually make safe supports for them.

Aside from considering the weight of your tank,

One of the most popular and hardy of the Lake Tanganyikan cichlids is *Cyphotilapia frontosa* originally described by Boulenger. There are many different morphs of this fish. Photo from A. Spreinat.

Even very large aquaria can be placed in most rooms of most houses, but precautions must be taken to ensure the safety of the fishes, the house, and the people living in the house.

Likewise, unless you are

you should take into account several other factors in determining where you put it. Avoid places of extreme temperatures—near outside doors, windows, or heat or air conditioning registers.

Another reason besides air temperature extremes to avoid locations near windows is that sunlight can be quite detrimental. Aquaria are great heat sinks, so on a sunny day, even in the winter, solar radiation can raise the temperature of the water considerably. In addition, excess sunlight will encourage algal growth, and while some Tanganyikan cichlids like to eat algae, it is hard to appreciate them if they're swimming around in pea soup.

Locate your Tanganyika tank where you will see and appreciate it often. It might look gorgeous in the living room, but if you spend most of your time in the den, you not only will miss a great deal of enjoyment from your fishes, you will also be less able to keep a watchful eye on them and be alert to conditions needing your attention.

Most Tanganyikan cichlids are curious and bold, and their tanks normally contain numerous hiding places, so a high-traffic area is fine, as long as the tank itself will not be subject to collisions, bumps, or excessive vibrations. As I write this my two preschoolers are playing a cross between hide-and-seek and jack-in-the-box with an empty carton right outside a 75-gallon rift lake tank whose inhabitants are behaving normally, alternating between absolute disinterest and mild curiosity in these activities on the other side of the glass.

Having your tanks handy is especially important when you are raising fry. The best way to get healthy adult fish is to feed the young tiny amounts many, many times a day. This is easy to accomplish when the tank is situated where you pass by it several times a day, and equally easy to neglect when the tank is out of sight (and out of mind).

The last consideration for tank location is its accessibility to electrical outlets, a water supply, and drain connections, and its ability to withstand the inevitable splashes and other water accidents. Even the wonderful suction/fill devices that many find so indispensable for water changes are not fool-proof, at least not for this fool, who has more times than he cares to admit left one of them inadequately secured only to turn his back and hear it clump out of the tank and onto the floor. It is absolutely amazing how much water can come out of that hose in the three seconds it takes to get to it, grab it, and flip the shut-off switch!

Speaking of filling your tank, if you're using a new aquarium once it is located in its final position you should fill it with cool tap water, then empty it completely, preferably after leaving the water in it for a day or two. This rinses out any dirt or harmful substances that might be in the tank, and the delay allows substances to leach out into the water. I had this lesson driven home to me once when I set up a new aquarium and got a lethal ammonia reading when I tested the water *before adding any fish!* It turns out that the warehouse the tank came from had resident pigeons, and residue from their droppings (which were scraped off before the tank left the warehouse) remained on the glass!

Temperature Regulation

I have already mentioned the temperature buffering of the rift lakes. Lake Tanganyika is the second largest body of fresh water on earth, and, situated in a very warm climate, it maintains at an almost constant, tropical temperature.

Do not skimp on the heater. There are numerous models to choose from, ones that hang on the rim as well as submersibles. You will need a quality thermostatically controlled aquarium heater of sufficient wattage to keep your tank at 80°F (26.5°C). If your home gets much warmer than this in the summer, you will either have to make provisions for cooling the aquarium water or else get an air conditioner for that room and enjoy the comfort along with your cichlids.

People often ask what size heater to get. Unfortunately, there is no simple answer to that question. An inadequate heater will, of course, fail to maintain the proper water temperature, causing your fishes not only not to thrive but even to fall sick and die. On the other hand, a heater of too many watts will not cycle on and off efficiently, and, should it get stuck in the on position, it will cook your fishes. There are two factors that affect the wattage needed.

The first factor is the ambient temperature; the closer a room's normal air temperature is to 80°F, the less powerful the aquarium heater must be. If the room normally is fifteen or more degrees cooler than the desired aquarium temperature, five watts per gallon may not even be sufficient. In addition, the more widely the normal room temperature fluctuates (from turning the heat down at night, for example), the greater the demands on the aquarium heater.

Secondly, the relative size of the aquarium is important. Fish tanks, like all solid objects, radiate heat as a function of their surface area. The heat *content* of a solid body, however, is a function of its volume. Now, area is a square measurement, while volume is a cubic one; hence, doubling the dimensions of a solid object (2d) quadruples the skin or surface area (2d x 2d =

Aquarium heaters are available in a wide range of lengths and wattages. It is a good idea to keep a spare heater on hand should one fail or burn out. Photo courtesy of Penn Plax.

$4d^2$), but *octuples* the body mass or volume (2d x 2d x 2d = $8d^3$). The differences quickly augment, and a tenfold increase in size (10d) gives a one hundred fold increase in surface area ($100d_2$), but a one thousand times increase in volume ($1000d_3$)!

Therefore, the bigger a solid body, the less heat loss it will have to the environment. The larger an aquarium, the less the relative heat loss, and the fewer watts per gallon needed. An aquarium 20 inches by 10 inches by 12 inches (10-gallon tank) has a combined surface area of its six faces of 1,120 square inches and a volume of 2,400 cubic inches. A tank 72 inches by 18 inches by 18 inches (100-gallon tank) has a total surface area of about 5,800 cubic inches and a volume of about 24,000 cubic inches. Thus, the larger tank has 10 times the volume, but only 5 times the area, so it loses heat half as quickly as the smaller tank, or, both to the considerable biological filtration they provide and to the microscopic organisms they foster—a domesticated replacement for the lake's natural microplankton. When used with a powerhead, sponge filters can provide ample biological filtration plus beneficial turbulence.

The major drawback of sponge filters is that to provide enough biological filtration for very large tanks the obligatorily large sponges are difficult to hide and not very attractive. As supplementary bio-filters, however, they are excellent, say in a tank that also has an undergravel system or a wet-dry filter.

Other bio-filters exist as well. Variations on the wet-dry theme are commonly available. They are quite different in design, from canisters full of odd-shaped plastic balls, through paddle wheel designs, to the newest fluidized bed filters, but all rely on the trickling, spraying, or soaking of a multi-layered, highly convoluted, or extremely porous medium, either flooding it with an oxygenated water flow, or letting water and air mix in the medium itself, the "wet-dry" approach. With these, air, which has many times the oxygen content of water, is present throughout the medium, increasing the number of aerobic bacteria greatly. These are extremely effective in removing ammonia.

One of the latest approaches really gets the bacteria out into the oxygen-rich air. It uses a rotating wheel of a pleated medium that is sprayed constantly with filtered water, producing an immense colony of bacteria in a very small space. This design is available in models that can connect to a variety of power filters and powerheads, making it possible to add this bio-filter to your existing filtration system. Because

Cyprichromis leptosoma. **Photo by MP & C Piednoir, Aqua Press.**

Lamprologus brichardi.

Lamprologus pulcher **"Daffodil."**

Lamprologus buescheri.

TAN(

Now th
all the ha
finally tim
"liveware"
you going
tank? In t
are going
several ge
Tanganyil
selected s
This will b
surface of
species, b
haphazard
are perfec
at a Tanga
are colorfu
breed, and
available. |
some of th
more exter
Tanganyik
can expan
your exper
initial spec
choose fisl
prove mos
you. Obvic
a source fc
you like th
mentioned
get informa
them, ther
you canno
those. This
only as a s
doorway w
enter the c
labyrinth c
vast numb
Tanganyik
species.

The Lamps:
Lamprologu

Certainly
most popul

One of the many morphs of *Tropheus moorii.* **Photo by Edward Taylor.**

the water is prefiltered, the bio-filter almost never needs cleaning, meaning the colonies of bacteria do not have to be disturbed.

Once established, the bacterial colonies must be cared for. The flow of oxygenated water over the medium must never be impeded or stopped, except for a few moments for servicing the filter. For an undergravel setup, regular vacuuming of the gravel surface to remove all trapped dirt is required. In bio-filters other than undergravel filters, the "cleaning" of the medium should consist only of an occasional rinsing in a bucket of aquarium water to remove particles that are

obstructing the water flow. Even then some bacteria may be lost, so it is a good idea to have more than one bio-filter and clean them alternately, or to clean just a portion of the medium at a time. Do not use chlorinated tap water to rinse bio-filter media, since the chlorine will kill the bacteria.

Mechanical Filtration

Mechanical filtration is the process by which suspended particles are screened out of the water, normally by passing a flow of aquarium water over a medium that is fibrous or sponge-like. The spaces in the medium trap the particles. By using varying

degrees of coarseness of material you can trap suspended particles from the largest to the tiniest. Many mechanical filters, especially canister filters, have a series of layers of successively finer filter media to progressively trap smaller and smaller particles.

The finest medium normally used is diatomaceous earth, the filter material in "water polishing" filters, which are designed to be used only on a temporary basis. These will filter out single-celled algae and even most disease organisms, but are not designed to serve as the primary mechanical filtration and run

plants, yo
course sel
artificial o
These plas
imitations
inedible a
indestruct
and they o
mind the v
conditions
pick live p
ask for tho
will tolerat
the water
chemistry
high
temperatu
the Tanga
tank, as w
the attacks
cichlids, s
Anubias or
Cryptocory
this case, y
also have t
consider th
lighting ve
carefully.

Lighting

Normal
lighting is e
for fishes t
eat and to

---

females. They do best in groups, with more females than males.

L. caudopunctatus. A pretty fish about two and a half inches long, which is starting to appear more regularly in dealers' tanks. Like many lamps, it has an understated beauty consisting mostly of iridescent highlights over a neutral-colored body, a beauty that is more easily seen in person than in photos.

(Alto)lamprologus compressiceps. This fish, along with its relative *calvus*, is sometimes placed in a separate genus, *Altolamprologus*, which means "deep *Lamprologus.*" Much deeper-bodied than other lamps, these fishes are often available in both light and dark forms. Their skinny profile enables them to enter narrow cracks to prey upon small fishes. Provide a lot of rock-work, since they mostly stay among the caves and crevices, where they spawn. They have relatively large clutches of 50 to 200 eggs, which the female guards.

L. sexfasciatus. A 3- to 5-inch, heftier lamp that is often available. The background can be either bluish silver or yellowish, with six dark vertical bands. They are cave-dwellers and cave-spawners.

L. tretocephalus. A 6-inch cichlid that differs from *L. sexfasciatus* by having one fewer black bar. The silvery

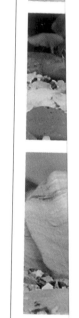

**A pair o
spawnin
The mal
shell, so
alone. T
opening
and squi
the eggs
Richter.**

body is
by an i
fins, es
Anothe
which
a coupl
is a pre

Cyprichromis nigripinnis, sometimes called the Blue Neon from Lake Tanganyika. Photo by Bob Allen.

Cyprichromis leptosoma from Karilani. Photo by MP & C Piednoir, Aqua Press.

Left:*Cyprichromis microlepidotus* from Kavala Islands. Photo by MP & C Piednoir, Aqua Press.

each egg in her mouth before it falls too far, and the male positions himself above the female and releases his milt as she delivers her eggs.

Cyprichromis are generally peaceful, as well as quite nervous and timid. They benefit from hiding places, such as among tall plants, but need plenty of open swimming space as well. These fishes should only be kept in large schools, which, along with

their active habits, require a large tank. They usually are not bothered by other, substrate-hugging, cichlids, and they provide visual interest in the upper part of the tank, which otherwise would normally be empty of activity.

Frontosa: *Cyphotilapia frontosa*

This fish rates a section of its own, since it is a popular aquarium fish as well as the only member of its genus. There is some variation by race, but the fish is basically bluish-silver with black vertical

A morph of *Cyphyotilapia frontosa*. Photo by Andreas Spreinat.

of cichlids that are less commonly available, but not overly difficult to find. Remember that this partial listing is morc like the tip than like the iceberg, and after you've gotten acquainted with Tanganyikan cichlids, you will always have plenty more species to try.

Other Species

The goby cichlids of the genera *Eretmodus*, *Spathodus*, and *Tanganicodus* are pair-forming mouthbrooders. By somc rcports they exchange the fry from mother to father part way through the brooding period. They are not extremely common in dealers' tanks, but they are worth looking for. These less than 3-inch fishes are

bars. The specific name refers to the large hump that develops on the forehead of the fish, especially on the males. This piscivorous cichlid congregates in great numbers at depths of 100 feet or more but is also found in shallow water.

It is lethargic for a predator, but fishes small enough to be swallowed are not safe in the same tank. They are usually peaceful for such a large cichlid, even at spawning time. *C. frontosa* is normally a reliable breeder, with clutches of about twenty eggs that the female mouthbroods for a very long time.

About the only drawback of this species is its size— up to 14 inches! Obviously they need large tanks and copious amounts of food, with live fishes the preferred dinner item.

The above species are the steady regulars of which you are likely to find a good assortment at your local dealer. I will next consider a few representative types

A closeup of the head of a male *Cyphotilapia frontosa*. Photo by Andreas Spreinat.

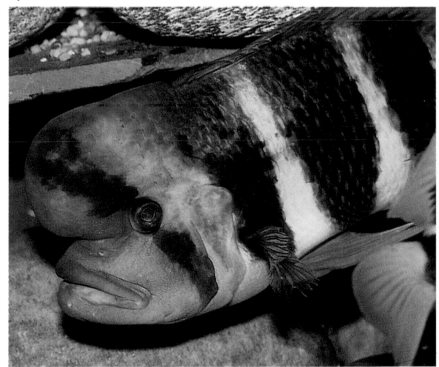

anal fin, which look like eggs, and the male releases sperm, which is taken into the female's mouth to fertilize the eggs. Males are quite territorial and combative but two males and several females can usually be maintained safely in a large aquarium.

You may find one or more species of the genus *Telmatochromis*. These small, slender fishes resemble species of *Lamprologus* and *Julidichromis*. They are notorious egg-stealers and their juveniles sometimes make use of their similarity in appearance to young *Julidichromis*, which tolerate their older offspring around a new spawn, to enter the school as wolves in sheep's clothing in order to devour the eggs.

Ophthalmotilapia ventralis ventralis. This morph is found on Chaitika Cape. Photo by Pierre Brichard.

Ophthalmotilapia ventralis. Photo by Ad Konings.

Ophthalmotilapia nasuta. Photo by Mark Smith.

FEEDING

Live Foods?

Older books on tropical fish keeping had extensive chapters on feeding, and they were necessary, since home-grown, home-collected, and home-made fish foods were required to be truly successful fish keepers and breeders. Today, however, the commercial market is replete with excellent choices of prepared, frozen, and freeze-dried foods in a variety that staggers the imagination. If you wish, you can make gourmets of your fishes! There are even foods specifically formulated for cichlids, and they are readily accepted by almost any species.

Live food is, of course, still of great benefit, but to be honest it is mostly psychological, that is, fishes seem to enjoy stalking and catching live prey. The breeding successes that still seem to hinge on live food cannot be accounted for in terms of nutritional analysis, but in some cases there does seem to be an edge with the live foods in terms of getting the fish "in the mood." Fortunately, you can have a very successful Tanganyika tank without ever feeding any live foods. I have raised tankfuls of fry that never saw a living dinner, except for whatever microorganisms lived naturally in their tanks.

Do not think I am against live food. I'm not, and I feed live foods to my own fishes. It's just that feeding live foods is optional, not necessary, most of the time. If you wish to feed live foods on a regular basis, your fishes will love you for it, and they will thrive. If you do not wish to do it, your fishes will also thrive, and probably still love you, at least at feeding time. The crucial thing is not whether your fish's diet is moving or not, but whether the diet is varied, nutritious, and complete. If you want to, you can confidently rely on the research and years of experience behind modern prepared fish foods, but older methods, so successful in the past, are still valid today.

One of the best live foods is that old favorite the brine shrimp, *Artemia salina*. Cichlid fry relish the newly-hatched shrimp, as do many adults of the smaller species, and this crustacean is an excellent substitute for the planktonic animals on which many Tanganyikans feed in nature.

Other fine live foods for your cichlids are *Daphnia*, *Gammarus*, and wingless *Drosophila*. Any source you have for these will also be able to provide you with culturing instructions, if

you also want to get into growing your own cultures. In my opinion, live *Tubifex* worms are not worth the risk of bacterial contamination, though you can buy bacteria-free freeze-dried *Tubifex*.

Other Foods

Even if you choose not to feed live foods, you must, of course, feed a varied, high quality diet, and you must pay attention to one specific dietary need—many Tanganyikan species are herbivorous, that is to say, adapted to an algal diet. They must be fed a good deal of plant material, either commercially prepared vegetable foods or blanched fresh veggies such as zucchini slices or spinach leaves. Even these fishes, however, benefit from a high protein supplement, especially when they are young and growing, and it is a rare Tanganyika cichlid that turns up its nose at animal protein.

Of course, food only does good if it is eaten, and without the instinctive attraction of living food items, prepared food has to appeal to your fish. It is rarely a problem to get Tanganyikan cichlids to eat, however, and they will do very well if you can feed them small quantities of many different high-quality foods numerous times a day.

A very easy supplemental food I have had good success with for all types of cichlids, from little Tanganyikan lamps all the way to adult South American oscars, is raw ground turkey. The tank literally boils with action when I toss the meat in and the feeding frenzy begins! Ground turkey has a distinct advantage over ground beef or beef heart, it is more easily digested and it contains much less fat. This is healthier for your fishes and for their tank, as beef often leaves a greasy film on the surface and chunks of undigested fat in the fish's feces. If you use turkey, make sure it is fresh—it can be frozen in one-use portions and defrosted before each use— and feed it sparingly.

Another food, which is a bit controversial due to its sometimes high concentration of pollutants, but which I have had great success with, is freeze-dried blood worms. What makes me recommend them is that fishes go crazy for them, and how they grow when they eat these high-protein insect larvae! When I have young cichlids I keep a canister of these delicacies on top of the tank so that I can sprinkle a few into the tank every time I go by as a supplement to their regular feedings. The fish respond eagerly to this regimen and grow very nicely. They also relish frozen bloodworms.

You should survey the foods available in your area and pick a varied diet both to round out your fishes' nutrition and to keep them from getting bored with a monotonous diet. There are almost as many "success diets" as there are successful fish breeders, but your fishes will do fine on a diet of high quality dried, freeze-dried, and frozen foods supplemented with vegetables and perhaps occasional live food and/or ground turkey.

Ophthalmotilapia nasuta. Photo by Hans Joachim Richter.

BREEDING

For many hobbyists, breeding their fish is the ultimate goal. Bringing your fish into condition, spawning them, and rearing the young to maturity is an indication that you are caring for them properly, and it is an exciting and fulfilling aspect of the hobby. In addition, it is often the way to expand your collection as you sell or trade your extra youngsters to get some new species. You won't get rich competing with professional breeding farms and importers, and you probably won't even cover the fish food bill, but you will be able to reinvest these assets in new specimens. This is another place where aquarium clubs can help, since they often have fish swaps or auctions associated with their meetings.

Breeding behaviors, like almost all features of Tanganyikan cichlids, evolved in response to the lake conditions. Since all of the fish's life is concentrated in relatively little space, some specialized care is necessary to ensure that at least some fry survive with all those hungry mouths around. Tanganyikan cichlids use one of two basic breeding strategies to provide that care— substrate spawning with vigorous defense of the young, or mouthbrooding.

Most substrate spawners lay their eggs in caves, often on the roof. Some, like species of *Altolamprologus*, utilize very narrow crevices just wider than the slim fish, that are easy to guard. Just secluding the eggs is not sufficient to ensure the survival of some fry, so these open spawners protect the young as well. In some species both parents guard their babies; in others only the mother does. Some, like certain shell-dwellers, form breeding colonies, in which they spawn as individual pairs but jointly defend the area against predators. Still others, like *Lamprologus brichardi*, defend breeding territories in which a single male spawns with several females; in this case even the youngsters help protect the younger fry.

There is usually not much difference, if any, between the sexes, but they can tell each other apart, so the time-honored method of procuring a group of juveniles and raising them together is recommended. In a batch of youngsters all the same age the largest will tend to be the males, but only tend to. It is often possible to sex the fishes by examining the genital papillae, but this stressful procedure is not for beginners.

You can raise some fry right in a community setting, but many fewer will survive compared to those in tanks set up for breeding. For the smallest shell-dwellers, a 10-gallon tank is spacious enough, and most of the small to medium sized cave-spawners can use a 15- or 20-gallon aquarium. Rock-work, flowerpots, PVC pipes, or some other materials should be used to make secluded breeding sites. Fine gravel is a good substrate, and many species will excavate elaborate caves under the rocks. When one of the fish—the female— disappears and refuses to leave a cave, it means that she probably has spawned. She fans the eggs, keeping a flow of oxygenated water passing over them. They will usually hatch in less than a week, and a few days after that the young will be free-swimming and looking for their first meal.

Various mouthbrooding strategies are utilized by Tanganyikan cichlids. There are both monogamous and polygamous mouthbrooders, and among the former we find species in which the parents both mouthbrood simultaneously, species in which they switch the eggs from one parent to the other, and species in which only the female cares for the eggs, which is also the case for the polygamous species.

These fishes hold their eggs in their throats, where they are constantly oxygenated by the parent's normal breathing—which draws water in through the mouth and expels it out through the gill openings. Their eggs can be larger in size than those of substrate spawners, since the increased care permits smaller clutches. Larger eggs mean that the fry can be larger, which means that they will be less likely to be eaten. Some species, like *Cyprichromis*, simply release their fry and swim away. Others, like *Tropheus*, continue to care for and protect the brood, taking them back into their mouth at any hint of danger. These species have the smallest clutches, since by the time the young leave their mother's mouth for the last time they are quite large.

In single-species breeding tanks, you can often leave the brooding

fish, and even the released young, in with the other members of the school. Many of the species that do not provide extended brooding nevertheless ignore the released fry. In any case, the baby fish are programmed instinctively to seek shelter from larger fishes, and in an aquarium with plenty of tiny hiding crevices, many of them will survive. If females carrying

Tropheus duboisi originally described by Marlier. This baby is going back into his mother's mouth. Photo by Bernd Melke.

eggs or young are unduly harassed by the other fishes in the aquarium, they should be carefully removed to nursery tanks. They should not be lifted from the water in the net, but transferred *under water* to a carrying container. If despite your precautions the brood is spit out, the mother may pick them back up again.

Some brooding fishes will eat small amounts of tiny food particles, and some will fast during the entire incubation period, but all

will be thin and weakened by the time they release the fry. Sometimes it is necessary to keep them in a separate tank for a while to get them back into condition before subjecting them to the stresses and competition of their regular tank and the advances of the males.

Feeding the Fry

Feeding baby fishes newly-hatched brine shrimp, the nauplii of *Artemia salina*, is an old and venerable tradition in the tropical fish hobby. The hatching of brine shrimp has become almost a sacred ritual, and the availability of this easily-produced fry food has for years made all the difference in countless spawning successes. Millions of fry have been raised on this almost-perfect food. For smaller fry than cichlids, hobbyists used to raise "infusoria," and smelly concoctions and brews were a necessary part of many a hatchery. Getting the fry to the size where they could take brine shrimp was the goal, since once the fish were on the shrimp, they almost always thrived. Thousands of aquarists still swear by these tiny crustaceans, which consistently produce spectacular growth results. Today, however, there are more options open to us for raising our Tanganyika spawns, since you can start them on live food or not.

Fishes still thrive on brine shrimp, of course, and there is absolutely nothing wrong with hatching and feeding this excellent live food. There is, however, no necessity of doing so. It is not uncommon for fry to survive unnoticed, thriving and growing while hiding out among the rocks in a Tanganyikan community tank. It's not the way to get large numbers of babies, but it demonstrates that these hardy little fishes are able to get their nourishment not only on the run from hungry, predacious adults, but also without the aquarist bothering to make special menus just for them. The fry survival rate, especially in the community tank, will improve greatly if you put baby brine shrimp into their hiding places, either with a meat baster or with a piece of tubing that you run down into their cave.

And even if you have the fry in a tank by themselves, this is one area where the psychological advantage of live food can be significant. Fry grow best when always full. They instinctively go for live foods. Live food organisms that are not eaten are usually still alive.

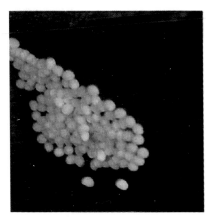 

Lamprologus leleupi is a magnificent fish and easy to spawn. The eggs are large and are attached to a rock in a secretive place. Once they are developed they must be plucked out of the egg cases. The egg cases are very tough as you can see from the cases still unopened. The fry stick together until their yolk sacs are absorbed and they are free swimming. Photos by Hans Joachim Richter.

All of this adds up to an easier maintenance with live brine shrimp nauplii, since the fishes will greedily hunt down the shrimp, which normally live long enough to get eaten, so there will be very few that die and start to decompose.

But you can also trade off where you want to spend time and money. If keeping batteries of shrimp hatcheries with their saltwater mixes, air stones, and collection challenges is a bother for you, you can use prepared food for your fry, taking care to feed tiny amounts extremely often, making sure you are even more assiduous with water changes and bottom siphoning, and keeping a close eye on the babies' girth to keep track of how much they are eating and experimenting with different foods until you can keep their bellies full.

Of course baby brine shrimp is also available already hatched, rinsed, and conveniently frozen. And there are any number of other suitable foods for newly-hatched cichlids. Still in the brine shrimp department, it is a simple matter to take a small piece of freeze-dried brine shrimp and pulverize it by rubbing it forcefully between thumb and forefinger. The technique requires only a little practice to produce an extremely fine-textured result. This powdered shrimp is eaten readily by baby cichlids.

The same method produces a powder-fine consistency from any high-quality flake food. Some manufacturers sell this powdered food, but it's just as easy to pinch and grind while you're feeding your older fishes as it is to open another container.

Before very long the fry are also able to nibble away at various wafers and pellets, some of which are available with 100% vegetable ingredients. These sink to the bottom and soften, where they are greedily attacked by the little fishes. Once your babies are able to feed this way, it is very easy to keep track of how much they are eating.

But whatever you feed them, feed the fry small amounts very often. I'll repeat that. Feed small amounts very often. Babies fed a variety of high-protein foods plus vegetables on a constant basis will grow quickly into healthy adults. Remember that the ideal is steady feeding to keep their bellies full; the more often you feed them, the better. This, of course, necessitates careful control of feeding portions, and almost constant water changes as well, but couple regular changes with good feeding and the only problem you should encounter is what to do with all the beautiful juveniles that will soon be crowding your rearing tanks.

The male *Tropheus* lies on his side while the female sucks the sperm into her mouth to fertilize her eggs. Photo by Bernd Melke. As a general rule, Tanganyikan cichlids are not difficult to spawn.

HEALTHY FISHES, HAPPY FISH OWNERS

It is much more important for you to understand disease prevention than for you to have a detailed knowledge of disease treatment, which is, in any case, beyond the scope of this book.

The treatment of fish diseases is a very complicated science, one not at all helped by the fact that our fishes cannot tell us where it hurts, the fact that many piscine ailments have similar symptoms, or by the distressing fact that many hobbyists first notice something is wrong when they find one or more fish belly-up in their tank.

In the event of a disease outbreak, all hope is not lost, but usually you must seek the advice of a competent professional, or of an experienced hobbyist, to diagnose and then treat your fishes. Unfortunately, many beginners skip the first two steps (advice and diagnosis) and immediately start dumping various medications into their aquarium. This is often useless, sometimes dangerous or lethal for the fishes, and only rarely successful.

The idea of a hobbyist well-experienced in treating fish diseases is a bit oxymoronic, since the most expert aquarists normally have little occasion to treat diseases. This is because they not only pay careful attention to their fishes' requirements, they also religiously follow two easy practices that ensure as far as is possible that their fishes will remain healthy. If you also follow these practices, without cutting corners, you will be able to avoid most problems:

Rule #1. *Caveat Emptor!*

The first and most important rule is to *buy healthy stock!* There are basically two ways of procuring your cichlids— from aquarium pet shops and from breeders or importers. You are certainly familiar with the first. Breeders and importers advertise in various hobby publications, and you can locate them through local aquarium societies and clubs as well. These sources often sell fishes for less than pet shops, but they often have minimum order sizes, and air shipping does not come cheap. So it evens out. Neither source for your fishes is better than the other, but as in any business transaction, if you are going to buy fishes from a price list, sight unseen, you should know something about the seller and get a written guarantee. There are many helpful, reliable, and trustworthy people in this business, and a little care will protect you from the few dishonest folk who try to ruin all walks of life.

Along the same vein, while most dealers are an excellent source and resource for aquarists, owning a pet shop does not make someone an honest, expert seller of only the finest quality fishes, so shop wisely. Even when you can inspect the fishes before buying, you should be careful. Know what to look for, and let wisdom, not passion, guide your decisions.

It is a terrible temptation, when you finally find a species that you've been aching to own, and you find it in a dealer's tanks at the right price, only a couple of the fish look sort of sick, and are those just a few spots on that one's body? Clamp your eyes shut and your hand over your wallet, grit your teeth, and walk away. Buying from this tank will only buy you trouble. Remember, no fish is worth risking your collection for.

Healthy cichlids are robust, active, with full (that is, not emaciated) bodies, but no swellings or protruding scales. They are alert and responsive, both to the other fishes in the tank and to humans outside the tank (either coming up to the front glass looking for food or fleeing quickly from the net). They have bright colors typical for that species, and they extend their fins fully and swim gracefully. Of course fishes with cottony fungus, white or red spots, lesions, or wounds should be avoided. Don't just look at the fish you plan to purchase. Check all the fish in the tank, and in other tanks in stores when there is a central filtration system that mixes the water from all the aquaria together.

Rule #2. Quarantine, Quarantine, Quarantine!

Quarantining new fishes is not an overprotective response of die-hard eccentrics. It is basic, intelligent fish husbandry. If you only own one tank, consider getting another one. If you have several, reserve one as a quarantine tank. It doesn't have to be big; it doesn't have to have any decorations; but it does need adequate filtration and heat.

It is best if the quarantine tank has an active biological filter. This can be accomplished (and the objection to having a totally empty and unused tank around can be answered) by maintaining a small population of some appreciated but expendable fish in it. These permanent residents not only serve to keep the bio-filter going and to keep the tank from looking useless, they also

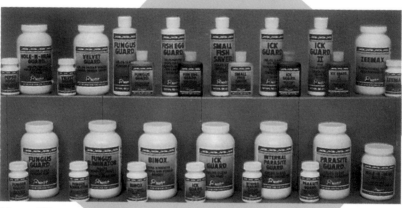

Many different remedies, preventatives and tonics are available at your local pet shop. Photo courtesy of Jungle Laboratories.

function as guinea pigs, as they will quickly alert you to any contagious conditions your new fishes might introduce.

If you choose to have permanent quarantine residents, or if you regularly acquire many new fishes so that the quarantine tank is often in use, you can include it as one of your display tanks. A quarantine aquarium does not have to be bare and sterile-looking. If you don't mind the possibility of dismantling the furnishings when it comes time to catch up the fishes to transfer them, you can have a regular substrate, rock formations, even plants. Of course, equally effective is a bare-bottomed aquarium

with a few PVC fittings for hiding places. Sponge filters are inexpensive and effective filters for such a tank.

Often overlooked is the benefit of quarantine on the newly-acquired fishes themselves. They are already stressed from being transported, sorted, purchased, and brought home. They have undergone pH shocks, hardness shocks, temperature shocks, food (or lack thereof) adjustments, as well as the stresses of being in typical sales situations of high population density and low or zero hiding places against aggressive tank mates. A bit of a rest in your care, acclimating to your water, your food, your schedule, will give them a head start before they have to deal with the stress of entering an established community.

How long to quarantine? There is no easy answer, except: the longer the better. Two weeks is a workable rule of thumb. In that time most pathologies will become evident. The more valuable the collection of fishes into which you are introducing the newcomers, the longer you should consider quarantining them.

It is not uncommon for healthy fishes to spawn in the quarantine tank! This is an added bonus and demonstrates both your practiced eye in selecting healthy, well-conditioned fishes and your success in acclimating them to your tank's conditions. In any event, by quarantining your new pets before you add them to your collection you will help ensure that your healthy tanks remain that way.

When Disease Strikes Anyway

If you do not buy obviously diseased fishes, if you quarantine them to make sure new fishes have no hidden diseases, and if you follow the guidelines for proper husbandry outlined in previous chapters, you will be nine-tenths of your way to eliminating disease by prevention.

That leaves those infrequent times when, despite your best precautions, your fishes become sick. Often this is due to some unavoidable problem, like a power outage and subsequent cooling of the water. An outbreak of ich, that white spot disease that is caused by a protozoan, often follows a chill. Your pet shop will have a choice of medications for this condition, and it is a good idea to have some on hand, just in case. Power failures

tend to occur during foul weather, and they seem to happen much more often when stores are closed!

Injuries, either from accidents or due to aggression from other fishes, require treatment. A commercial preparation can be used on injured fishes to forestall fungus, something definitely easier to prevent than to treat. It is preferable to treat just the affected individual in a small isolation tank with water taken from the main tank to avoid more shock to its system.

Internal parasites can be a problem, especially in

wild-caught fishes. Affected fishes have whitish feces and may regurgitate their food. Some cichlid hobbyists routinely treat new purchases with a vermicide. There is a flake food available with a worming medication in it. This has the benefit of concentrating the medication where it is needed and should be fed according to label instructions.

For any other problems, you should consult either a good book on fish diseases, an advanced hobbyist, or a piscine veterinarian.

Altolamprologus calvus **in its yellow morph. Photo by MP & C Piednoir, Aqua Press.**

TANGANYIKA TIME

Armed with the information in this book, you are now prepared to create your own Tanganyika microcosm, bringing the wonders of the diverse fauna of this African lake into your home. With their water requirements met, the beautiful and fascinating cichlids of Tanganyika are undemanding and hardy, and their readiness to breed in captivity means that you will be able to experience the marvels of the unusual parental care exercised by these fishes.

As you continue in the hobby, you may want to specialize, or, like many aquarists, you may enjoy an increasing assortment of species, perhaps adding species from Lake Malawi, whose cichlids responded in slightly different ways to similar evolutionary pressures, to your collection. In any event, you are unlikely ever to become bored with the wealth of fish species, and it is also doubtful that even after many years you will cease to marvel at the beautiful diversity of your Tanganyika tank.

Not all of Lake Tanganyika are rocky habitats. There is a reed zone and there are many small fishes to be found in this zone. Photo by Hans Joachim Richter